Plants of the Falkland Islands

Ali Liddle

FALKLANDS CONSERVATION

First published in Great Britain in 2007 by Falklands Conservation

ISBN: 978-0-9538371-9-9

United Kingdom OTEP Overseas Territories Environment Programme

This book has been produced with support from the
UK Overseas Territories Environment Programme

Falklands Conservation:
1 Princes Avenue, Finchley, London N3 2DA, UK

and

PO Box 26, Stanley, Falkland Islands.

www.falklandsconservation.com

Designed by Eye-on-the-Ball Ltd

Printed by ABC Print, Hampshire

Falklands Conservation is a UK registered charity no. 1073954 and company limited by guarantee in England and Wales no. 3661322 and registered as an overseas company in the Falkland Islands.

This book is dedicated to you, Mum.
Thank you for everything.

CONTENTS

Acknowledgements

Firstly, I am grateful to Alan Henry for the use of his many excellent photographs and for his enthusiasm, which enabled me to have to the courage to get this book off the ground in the first place. Many thanks also to Robin Woods for his expert advice and help with editing, and to Ann Brown for her keen interest and work getting the book into print. I gratefully acknowledge the important contribution made to the content of this book by Jim McAdam, the late Tom Davies, Robin Woods, David Broughton and Professor David Moore whose publications have been an invaluable source of botanic information. Finally thanks to my Mum, Norma Herring, for accompanying me on my many photographic outings around the Falkland Islands, not always in fine weather!

Photo Credits

All images in this book courtesy of the author Ali Liddle except for the following, courtesy of Alan Henry: page 29 (MAIN), page 31 (INSET), page 35 (MAIN), page 36 (ALL), page 37 (LEFT), page 38 (MAIN), page 39 (INSET), page 40 (MAIN), page 41 (MAIN), page 42 (MAIN), page 43 (MAIN), page 44 (MAIN), page 48, page 52 (MAIN), page 55 (MAIN), page 56 (ALL), page 57 (RIGHT), page 58, page 60 (ALL), page 62, page 64 (ALL), page 65 (MAIN), page 66 (MAIN), page 67, page 68 (MAIN), page 77 (RIGHT), page 78 (RIGHT), page 79 (RIGHT), page 81 (LEFT), page 83 (BOTTOM LEFT), pages 88/89 (BOTTOM) and AG Jones page 14 (TOP LEFT).

Introduction

At first glance the Falkland Islands may appear to have a uniform and uninteresting landscape with vegetation limited to straw-coloured white grass, which dominates vast areas. Closer inspection will be rewarded by the discovery of a beautiful range of plants. This book aims to help with the identification of many of the species and also includes information on habitats and some local recipes.

Fifty six of the more common and attractive plants are described here out of a total Falklands flora of 348 recorded species, of which 171 are native to the Islands and 13 are endemics which do not occur naturally anywhere else in the world. Due to the proximity of the Islands to South America, many Falklands plants are also found in the southern parts of the South American continent, throughout Patagonia and Tierra del Fuego. The remaining 177 species have been introduced over many years, some deliberately others accidentally. Most plants are widespread throughout the archipelago, but today Beauchêne Island is the only place which remains free of non-native species.

Tussac grass is a particularly important plant for the Falklands. In a land without trees and with persistent strong winds, over three quarters of Falkland birds use this habitat for feeding and nesting. It also provides shelter for seals and is favoured by many of the Island's invertebrates. Over 80% of tussac grass has disappeared in the last few hundred years due to grazing by sheep and cattle. Despite this decline, what is left is thought to account for a considerable proportion of the world population of this plant.

Given the low density of human population and inaccessibility to many areas, there are still places where little is known about the distribution and densities of Falkland plants. We still have a lot to learn about their distribution and ecology. Falklands Conservation would be very pleased to hear from anyone who has noted and/or photographed unusual wild plants or plant communities. If you have a particular interest or query you might like to visit the Falklands National Herbarium in Stanley, run by Falklands Conservation, which houses reference specimens of most plants including many seaweeds.

Today, with rapid developments in both agriculture and tourism, combined with new roads to remote areas, potential threats to the Islands' special plants continue. We hope that this book will go some way to stimulating a greater interest, knowledge and appreciation of the Islands' flora.

FACING PAGE: *Pale maiden, the national flower of the Falkland Islands.*

Special Plants of the Falkland Islands

There are 13 plants found in the Falkland Islands and nowhere else in the world. Some of these endemic species (E) are included in the 29 plants legally protected (P) under the 1999 Conservation of Wildlife and Nature Ordinance, which makes it an offence to deliberately pick, collect, cut, uproot or destroy them. 23 of the Falkland plants included in the IUCN (The World Conservation Union) Red List (R) are nationally threatened. Some of these species have very limited locations and are vulnerable to disturbance. All these special plants are listed in the table below.

ENGLISH NAME	SCIENTIFIC NAME	STATUS				PAGE
Adder's tongue	*Ophioglossum crotalophoroides*		P	R	Vulnerable	
Antarctic cudweed	*Gamochaeta antarctica*	E	P	R	Endangered	
Chilean tall fern	*Blechnum cordatum*		P	R	Vulnerable	
Clubmoss cudweed	*Chevreulia lycopodioides*	E				
Coastal naussauvia	*Nassauvia gaudichaudii*	E				(p43)
Common violet	*Viola maculata*		P			(p69)
Dusen's moonwort	*Botrychium dusenii*		P	R	Vulnerable	
Falkland smooth ragwort	*Senecio vaginatus*	E				(p44)
Falkland woolly ragwort	*Senecio littoralis*	E				(p45)
False-plantain	*Nastanthus falklandicus*		P	R	Vulnerable	
Fir clubmoss	*Huperzia fuegiana*		P	R	Endangered	
Fuegian violet	*Viola magellanica*		P	R	Vulnerable	
Fuegian whitlowgrass	*Draba magellanica*		P	R	Critically Endangered	
Gaudichaud's orchid	*Chloraea gaudichaudii*		P			(p56)
Hairy daisy	*Erigeron incertus*	E	P	R	Vulnerable	

Lady's slipper	*Calceolaria fothergillii*	E				(p 48)
Leathery shield-fern	*Rumohra adiantiformis*		P	R	Endangered	
Maidenhair fern	*Adiantum chilense*		P	R	Endangered	
Moore's plantain	*Plantago moorei*	E	P	R	Vulnerable	
Mudwort	*Limosella australis*		P	R	Vulnerable	
Pale yellow orchid	*Gavilea australis*		P	R	Vulnerable	(p 57)
Patagonian hawkweed	*Hieraceum patagonicum*		P	R	Endangered	
Pondweed	*Potamogeton linguatus*		P			
Rock-cress	*Phlebolobium maclovianum*	E	P	R	Vulnerable	
Saxifrage	*Saxifraga magellanica*		P	R	Critically Endangered	
Shrubby seablite	*Suaeda argentinensis*		P	R	Critically Endangered	
Silvery buttercup	*Hamadryas argentea*	E				
Skullcap	*Scutellaria nummulariifolia*		P	R	Critically Endangered	
Snake plant	*Nassauvia serpens*	E				
Spider flower	*Arachnitis quetrihuensis*		P	R	Vulnerable	
Tasselweed	*Ruppia filifolia*		P	R	Vulnerable	
Vanilla daisy	*Leucheria suaveolens*	E				(p 37)
Yellow lady's slipper	*Calceolaria biflora*		P	R	Critically Endangered	(p 48)
Yellow pale maiden	*Sisyrinchium chilense*		P			(p 53)
Yellow orchid	*Gavilea littoralis*		P			

Habitats of the Falkland Islands

The Falklands environment is influenced by its strong and persistent winds, which have resulted in a generally low-growing native vegetation. Introduced trees are restricted to sheltered gardens in Stanley and around settlements. The height plants attain can vary greatly. Wild celery might be a ground cover plant on sandy coastal greens but amongst the shelter of tussac grass grows to over 40 cm. Arrow-leaved marigold can survive as a tiny plant, virtually flat to the ground, but by sheltered ponds and pools flourishes with long stems and large succulent leaves. The tallest native plants are tussac grass, fachine and boxwood, which can grow up to two metres high. On ungrazed offshore islands, tussac grass may occasionally reach three metres.

The different habitats described are influenced by soil type, moisture and shelter. Each has its own particular range of plants, though many are not restricted to just one type. Diddle-dee, for example, is found both on coastal heathland as well as inland mountain slopes. The plant habitat symbols are used throughout the book to indicate a plant's most common location.

COASTAL MARGINS, MARITIME ROCK, SHINGLE, CLIFF and SLOPE

This includes rocky areas, shingle and cliffs influenced by wave-splash/sea-spray with only a few plant species such as sheep's sorrel, wild celery, thrift, and cushion plants such as balsam bog.

Rocky coast

Cliff

SAND DUNES

These are areas often covered in marram grass, with other plants such as sea cabbage, stonecrop and sheep's-sorrel.

Sand dunes

TUSSAC

This giant grass is usually confined to coastal regions less than 200 metres above sea level, particularly on small offshore islands and protected headlands. Other plants may grow in small clearings between the pedestals. Wild celery and Falkland strawberry are commonly found. Tussac grassland habitat was once much more widespread in the Islands and has been dramatically reduced as a result of grazing practices.

Tussac

INLAND HABITAT TYPES

These are dominated by a variety of heathlands on generally peaty soils and are usually found above the tussac grass line, often extending to the higest peaks.

ACID GRASSLAND

Grassland dominated by white grass, with short rush in wetter regions. White grass covers vast areas, particularly in Lafonia. Where few or no sheep graze, it provides vital shelter to many inconspicuous and attractive small plants. Acid grassland is generally found on poorly drained peat.

DWARF SHRUB HEATH

This habitat is generally found on better-drained shallower soils. It is dominated by diddle-dee and other common species including mountain berry and Christmas bush.

FERN BEDS

Patches of tall fern found within dwarf shrub regions.

SCRUB

Dominated by fachine, boxwood or the introduced gorse. Small stands of native shrubs are often remnants of much larger areas, which have been reduced by intensive grazing.

Acid grassland

Dwarf shrub heath

Scrub

FELDMARK

Stony or clay soil areas which support cushion plants, mosses and lichens (generally over 500 metres above sea level, though it can be found nearer sea level in exposed areas). There are often patches of bare soil.

INLAND ROCK

These are inland rocky outcrops, cliff faces and stone runs which are virtually devoid of vegetation except for lichens. In sheltered crevices and between stone run boulders a surprising variety of plants can be found.

Feldmark

Inland rock

WET HABITATS

This includes rivers, streams, ponds and lakes as well as wet areas of ground.

NEUTRAL GRASSLAND (GREENS)
These are unimproved moist/waterlogged grasslands on neutral soils. Often heavily grazed, coastal greens behind the high tide mark contain introduced grasses and low-growing plants including rushes and cinnamon grass.

FEN, MARSH AND SWAMP
Waterlogged and wet areas often bordering streams, with plants including native rush and pig vine. NOTE: this does not include waterlogged grasslands (see Neutral Grasslands above)

BOG
Peat-forming vegetation on poor wetlands, dominated by low-lying plants including astelia and sundew.

Coastal greens

Wet habitats

NON-NATIVE HABITATS

There are close to 80 settlements spread across the Falkland Islands with gardens containing many introduced plants. Much of the native vegetation has been modified in many places by farming. Improved grasslands have replaced heathland in many localities. These have been created by burning to remove native plants, the sowing of grasses for fodder and/or the application of fertilisers and herbicides. There are some fields of arable crops/commercial horticultural land and several conifer plantations. Gorse hedgerows, introduced from Europe to keep in cattle and sheep, often dominate the landscape.

BUILT UP AREAS / GARDENS

In Stanley, introduced plants form many of the hedges and there are some fine examples of Monterey cypress, particularly in the older part of town.

WOODLAND

Plantations of mature conifers where broad-leaved trees make up less than 20% of the woodland. This includes glades or rides within the woodland.

IMPROVED GRASSLAND

These areas have few plant species except grasses, which have been either sown or modified by fertilisers and/or herbicides.

Stanley

Settlements

How to use this book

Scope of book

This book contains many of the common Falkland plants, and a few which are less common but selected because they are particularly attractive or interesting. Plants have been included from nearly all Falkland habitat types and from most of the plant families found growing in the Islands.

KEY TO THE SPECIES

On pages 18–23 is a key to help you identify the plants.

PLANT DESCRIPTIONS

Order of plants

The plants are ordered as follows: Flowering plants, Trees, Grasses, Rushes and Ferns. Within this, they are listed alphabetically by family, with similar species *e.g.* hawkweeds grouped together.

The descriptions

Each species is illustrated by a main photograph and, where useful, additional photographs to aid identification.

Icons indicate the habitat in which a species occurs (*see key below*). Its status within the Falkland Islands is shown in a coloured box.

The text covers a brief overview of the plants and a description of, where appropriate, the flower, leaf, stem and fruit/seed. The period during which the plant is in flower and an indication of places to look are also given.

There is also a separate paragraph covering local information, including reference to the pages where recipes using the plant can be found.

Key to habitat icons

Inland rock

Rocky coast

Acid grassland

Coastal greens

Cliff

Dwarf shrub heath

Wet habitats

Sand dune

Feldmark

Settlements

Tussac

Scrub

Stanley

Key to the plants

This key covers all the plants illustrated in this book. Basic details of each plant's form covering flower colour, flower shape, plant height/spread and fruit are given along with details of flowering/fruiting period and habitat.

FLOWERING PLANTS

ENGLISH NAME	SCIENTIFIC NAME	COLOUR	FLOWER SHAPE
Balsam Bog	*Bolax gummifera*	white	
Wild Celery	*Apium australe*	white	
Mountain Berry	*Pernettya pumila*	white	
Almond Flower	*Luzuriaga marginata*	white	
Field Mouse-ear	*Cerastium arvense*	white	
Dusty Miller	*Primula magellanica*	white	
Dog Orchid	*Codonorchis lessonii*	white	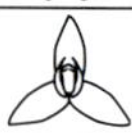
Sundew	*Drosera uniflora*	white with yellow-green centre	
Falkland Strawberry	*Rubus geoides*	white with yellow-green centre	
Teaberry	*Myrteola nummularia*	white with pink centre	
Pale Maiden	*Olsynium filifolium*	white with yellow centre	
Fachine	*Chiliotrichum diffusum*	white with yellow centre	
Marsh Daisy	*Aster vahlii*	white with yellow centre	
Vanilla Daisy	*Leucheria suaveolens*	white with pale golden centre	
Stonecrop	*Crassula moschata*	white or pink	
Boxwood	*Hebe elliptica*	white or lilac	
Scurvy Grass	*Oxalis enneaphylla*	white, mauve to pale pink	
Berry-lobelia	*Pratia repens*	pale lilac with yellow centre	

KEY 🟠 in flower 🔴 in fruit 🟢 spores present

FLOWER SIZE: flower diameter categories
tiny: less than 6 mm **small**: 6–12 mm **medium**: 12–20 mm
large: 20–40 mm **very large**: larger than 40 mm

FLOWER SIZE	HEIGHT/SPREAD	FRUIT	IN FLOWER/FRUIT S	O	N	D	J	F	M	A	M	PREFERRED HABITAT	PAGE
tiny	up to 1·5 m × 1·5 m; cushion			🟠	🟠							dwarf shrub heath, cliff, feldmark	**30**
tiny; clustered	15–80 cm				🟠	🟠	🟠	🟠	🟠			coastal greens, tussac	**31**
small	up to 4 cm; spread to 50 cm	deep pink			🟠	🟠			🔴			dwarf shrub heath, inland rock	**51**
medium	up to 4 cm; creeping	dark red			🟠	🟠	🟠					dwarf shrub heath, inland rock	**60**
medium	up to 5 cm; spread to 25 cm				🟠	🟠	🟠	🟠				dwarf shrub heath, acid grassland	**61**
medium; clustered	10–20 cm			🟠	🟠	🟠						dwarf shrub heath	**62**
medium	10–25 cm				🟠	🟠						dwarf shrub heath, acid grassland	**55**
small	up to 3 cm						🟠	🟠				wet habitats	**67**
medium	up to 2 cm; ground cover	red; cluster			🟠	🟠	🔴					acid grassland, tussac	**64**
small	up to 3 cm; spread to 15 cm	pink and white			🟠	🟠	🟠		🔴	🔴		acid grassland. dwarf shrub heath	**54**
large	20 cm				🟠	🟠	🟠	🟠				dwarf shrub heath	**52**
large	up to 2·0 m × 2·0 m					🟠	🟠	🟠				scrub dwarf shrub heath	**38**
medium to large	30 cm						🟠	🟠				acid grassland	**36**
very large	up to 25 cm				🟠	🟠	🟠					acid grassland, dwarf shrub heath	**37**
tiny	up to 2 cm; creeping					🟠	🟠	🟠				sand dune. rocky coast	**66**
medium	up to 2·5 m × 2·0 m; shrub					🟠	🟠	🟠				Stanley, cliff	**47**
large	up to 20 cm			🟠	🟠	🟠	🟠	🟠				dwarf shrub heath, acid grassland	**71**
medium	up to 2 cm; creeping					🟠	🟠	🟠				sand dune, open water	**29**

ENGLISH NAME	SCIENTIFIC NAME	COLOUR	FLOWER SHAPE
Felwort	*Gentianella magellanica*	lilac	
Falkland Lavender	*Perezia recurvata*	lilac, pale blue or white	
Pimpernel	*Anagallis alternifolia*	pale pink	
Falkland Thrift	*Armeria macloviana*	rose pink	
Astelia (Soft-camp-bog)	*Astelia pumila*	cream	
Christmas Bush	*Baccharis magellanica*	cream	
Coastal Nassauvia	*Nassauvia guadichaudii*	cream	
Pale Yellow Orchid	*Gavilea australis*	cream with pale yellow centre	
Gaudichaud's Orchid	*Chloraea gaudichaudii*	pale green with white centre	
Bead Plant	*Nerteras granadensis*	pale green	
Arrow-leaved Marigold	*Caltha sagittata*	pale yellow with yellow-green centre	
Buttonweed	*Leptinella scariosa*	yellow	
Sea Cabbage	*Senecio candidans*	yellow	
Groundsel	*Senecio vulgaris*	yellow	
Antarctic Hawkweed	*Hieracium antarcticum*	yellow	
Falkland Smooth Ragwort	*Senecio vaginatus*	yellow	
Falkland Woolly Ragwort	*Senecio littoralis*	yellow	
Yellow Pale Maiden	*Sisyrinchium chilense*	yellow	
Gorse	*Ulex europaeus*	yellow	
Yellow Lady's Slipper	*Calceolaria biflora*	yellow	

FLOWER SIZE	HEIGHT/SPREAD	FRUIT	IN FLOWER/FRUIT									PREFERRED HABITAT	PAGE
			S	O	N	D	J	F	M	A	M		
medium	12 cm						●	●				dwarf shrub heath	49
large	up to 30 cm; spread to 1 m					●	●					dwarf shrub heath	42
small	up to 2 cm; ground cover					●	●	●				sand dune, cliff	63
tiny; clustered	up to 15 cm				●	●	●					cliff, sand dune	68
small	up to 2 cm; cushion				●	●	●					acid grassland	26
small	up to 8 cm; spread to 1 m					●	●					dwarf shrub heath, scrub	34
tiny	up to 10 cm × 1 m; cushion					●	●	●				sand dune. dwarf shrub heath	43
large	up to 25 cm					●	●					dwarf shrub heath	57
large	10–25 cm				●	●						dwarf shrub heath, acid grassland	56
tiny	up to 2 cm; ground cover	bright orange red		●	●	●	●	●	●	●		sand dune acid grassland	27
medium to large	up to 20 cm			●	●	●						wet habitats	28
small	up to 2 cm; ground cover					●	●					sand dune, wet habitats	33
medium	up to 1 m					●	●	●				sand dune. coastal greens	35
tiny	up to 40 cm; bushy		●	●	●	●	●	●	●	●	●	dwarf shrub heath	39
medium	30–40 cm					●	●	●				Stanley, sand dune	40
large	up to 20 cm				●	●	●	●				dwarf shrub heath	44
large	up to 25 cm; bushy				●	●	●					dwarf shrub heath	45
medium	15 cm					●	●	●				coastal greens, acid grassland	53
medium	up to 2 m × 10 m; shrub			●	●	●	●					settlements, Stanley, scrub	59
large	up to 12 cm				●	●	●	●	●			dwarf shrub heath	48

FLOWERING PLANTS

ENGLISH NAME	SCIENTIFIC NAME	COLOUR	FLOWER SHAPE
Common Violet	*Viola maculata*	yellow	
Yellow Orchid	*Gavilea littoralis*	deep yellow	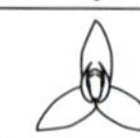
Lady's Slipper	*Calceolaria fothergillii*	orange	
Orange Hawkweed	*Hieracium aurantiacum*	deep orange	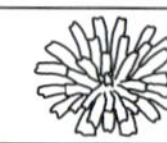
Sheep's Sorrel	*Rumex acetosella*	red	
Diddle-dee	*Empetrum rubrum*	MALE: deep red FEMALE: chestnut	
Pig Vine	*Gunnera magellanica*	MALE: red FEMALE: deep pink	FEMALE 
Prickly Burr	*Acaena magellanica*	reddish-green	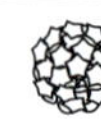
Water-milfoil	*Myriophyllum quitense*	purplish green	

TREES

ENGLISH NAME	SCIENTIFIC NAME	
Monterey Cypress	*Cupressus macrocarpa*	

GRASSES & RUSHES

ENGLISH NAME	SCIENTIFIC NAME	LEAVES	HEIGHT/SPREAD
Cinnamon Grass	*Hierochloe redolens*	green; shiny; broad; cinnamon aroma when crushed	up to 70 cm; clumps
Marram Grass	*Ammophila arenaria*	green; shiny; thin	up to 1·5 m; clumps
Tussac Grass	*Poa flabellata*	green; thin; very long	up to 3 m; clumps
White Grass	*Cortaderia pilosa*	sandy-brown; wiry	up to 40 cm; tussocks/individual
Native Woodrush	*Luzula alopecurus*	red-brown to light green	15 cm
Short Rush	*Rostkovia magellanica*	brown and green; slender	20 cm
Tall Rush	*Marsippospermum grandiflorum*	dark green and reddish-brown; slender	35 cm

FERNS

ENGLISH NAME	SCIENTIFIC NAME	FRONDS	HEIGHT
Tall Fern	*Blechnum magellanicum*	YOUNG: red-brown and bright green; MATURE: dark green	up to 70 cm
Small Fern	*Blechnum penna-marina*	YOUNG:bronze and pale green MATURE: dark green	up to 15 cm

FLOWER SIZE	HEIGHT/SPREAD	FRUIT	S	O	N	D	J	F	M	A	M	HABITATS	PAGE
			IN FLOWER/FRUIT										
medium	up to 15 cm				●	●	●					dwarf shrub heath, sand dune	**69**
large	20–40 cm				●	●	●					cliff, dwarf shrub heath	**58**
large	up to 12 cm				●	●						dwarf shrub heath, sand dune	**48**
large	up to 30 cm					●	●	●	●			Stanley	**41**
tiny	up to 20 cm; ground cover				●	●	●	●				sand dune, dwarf shrub heath	**46**
tiny	10–50 cm; ground cover	deep crimson to bright orange-red		●			●(red)					dwarf shrub heath	**32**
tiny	5–20 cm; ground cover	bright orange-red; cluster		●	●	●		●(red)	●(red)	●(red)		dwarf shrub heath, acid grassland	**50**
medium to large	20–40 cm				●	●	●	●				sand dune	**65**
tiny	aquatic; depth dependent						●	●				open water	**70**

	HEIGHT/SPREAD	FRUIT	S	O	N	D	J	F	M	A	M	HABITATS	
			IN FRUIT										
	10–20 m	young cones: green; mature cones: brown	●	●	●	●	●	●	●	●	●	settlements, Stanley	**72**

FLOWER	S	O	N	D	J	F	M	A	M	HABITATS	PAGE
	IN FLOWER										
shiny green; clustered	●	●	●	●	●					sand dune, dwarf shrub heath	**73**
green; narrow				●	●	●				sand dune	**74**
green; oval; large	●	●								tussac	**75**
sandy-brown; delicate				●	●					acid grassland	**76**
pale-brown with fluffy white hairs; egg-shaped			●	●	●					dwarf shrub heath, inland rock	**77**
FLOWER: insignificant; SEED: distinctive brown, round capsule		●	●							acid grassland, wet habitats	**78**
FLOWER: brown; crown-shaped		●	●							dwarf shrub heath, acid grassland	**79**

	S	O	N	D	J	F	M	A	M	HABITATS	PAGE
	SPORES PRESENT										
							●	●		dwarf shrub heath	80
							●			dwarf shrub heath, acid grassland	81

GRAND JASON
STEEPLE JASON
FLAT JASON
SEDGE ISLAND
JASON ISLANDS
ELEPHANT JASON
SOUTH JASON
SAUNDERS ISLAND
KEPPEL ISLAND
PEBBL
CARCASS ISLAND
DUNBAR ISLAND
WEST POINT ISLAND
Dunbar
Byron Sound
Shallow Bay
WEST FALKLAND
SPLIT ISLAND
Hill Cove
MT CAROL
Roy Cove
MT ADAM
Crooked Inlet
2297
King George Bay
PASSAGE ISLANDS
Chartres
NEW ISLAND
Dunnose Head
SWAN ISLAND
Queen Charlotte Bay
Spring Point
Lake Sulivan
BEAVER ISLAND
Port Richards
Fox Bay West
Fox Bay East
GREAT ISLAND
Smiley Channel
Lake Hammond
WEDDELL ISLAND
Double Creek
South Harbour
Port Edgar
RUGGLES ISLAND
ELEPHANT CAYS
Port Stephens
Hoste Inlet
BIRD ISLAND
Port Albermarle
SPEEDWELL ISLAND
ARCH ISLANDS
Cape Meredith
GEORGE ISLAND
BARRE
1500
1000
500
0
Heights in feet above sea level
61°W
60°W

THE FALKLAND ISLANDS

Astelia family

Astelia (Soft-camp-bog)

Astelia pumila

NATIVE

This is a low-growing ground cover plant, which forms hard cushions on wet, peaty soil. It reaches no more than 2 cm in height. The leaves are very stiff and sharply pointed with a deep groove along the centre and have a shiny upper surface. They vary in colour from dark green to bright, almost lime green. The small flowers are cream with a silvery coloured base (INSET TOP).

IN FLOWER: November to January.

PLACES TO LOOK: Widespread but most commonly found amongst white grass on very wet, peaty ground locally known as 'soft camp' (INSET BOTTOM).

LOCAL INFORMATION: When driving or even riding a horse across camp it is always wise to avoid areas of astelia as it grows on notoriously wet ground (INSET) where the water table may be right at the surface. If the top layer of vegetation is broken a serious 'bogging' is likely to occur!

Bedstraw family

Bead Plant

Nertera granadensis

NATIVE

A low-growing creeping ground cover plant that can form small carpets of colour amongst white grass. It rarely exceeds 2 cm in height but the stems can creep over 40 cm. The leaves are pale yellow-green, heart-shaped and grow along the length of the woody stems. The tiny pale green flowers are almost invisible but its perfectly round and bright orange-red berries are very conspicuous.

IN FLOWER: October and November, with berries from December onwards.

PLACES TO LOOK: Damp peat areas near the coast and amongst wet areas of white grass.

LOCAL INFORMATION: Bead plant can go unnoticed amongst other vegetation until its berries ripen. Thankfully they often remain on the plant throughout the winter allowing for easy identification.

Arrow-leaved Marigold

Caltha sagittata

NATIVE

This is a low-growing plant with distinctive shiny, green, arrow-shaped leaves. Height varies from 2–20 cm depending on location. The stems are broad woody rhizomes that often appear above the soil surface. It has pale yellow flowers with quite robust petals and a large yellow-green centre. The petals have very obvious green veins leading to the centre of the flower. It produces fantastic spiky seed heads containing up to 70 seeds each (INSET).

IN FLOWER: October to December.

PLACES TO LOOK: Wet, peaty areas by ponds and streams and sometimes in short grass in coastal areas.

LOCAL INFORMATION: This is one of the first flowers to emerge in spring. Its size varies greatly depending on the habitat, from a tiny flat ground cover plant in short grass to a large, succulent plant by ponds.

Berry-lobelia

Pratia repens

NATIVE

A small, low-growing, creeping plant that forms dense mats in damp, coastal areas. It is no more than 2 cm in height but the creeping stems can be up to 7 cm in length. The dark green leaves are round, quite succulent and often difficult to see amongst taller vegetation. The flowers (INSET) are very pale lilac, almost white, with a yellow centre. They are particularly distinctive with five petals offset to one side leaving a very obvious gap on the other, looking as if someone has pulled the petals off one side!

IN FLOWER: December to February.

PLACES TO LOOK: Damp, coastal greens and around fresh water ponds.

LOCAL INFORMATION: This plant remains inconspicuous until summer when the flowers appear and create delicate 'star speckled' carpets of colour.

Carrot family

Balsam Bog

Bolax gummifera

NATIVE

Balsam bog is a large cushion plant, which is surprisingly hard to the touch. Its rounded cushions can be up to 1·5 m in height, and often reaches a diameter of 1·5 m. The outer cushion is made up of tightly packed stems covered with small pointed leaves. This grows on an inner cushion of decaying leaves from previous years' growth. Tiny white flowers appear on the surface of the bog. In summer the cushions ooze drops of sticky, white, fragrant latex (INSET) which dries brown and hard during the winter.

IN FLOWER: October to November.

PLACES TO LOOK: Dry, rocky coastal heathland and mountainous areas.

LOCAL INFORMATION: The white resin has medicinal properties and was used in the past to heal cuts and sores. This resin is strongly aromatic, particularly on warm, calm summer days when entire areas of coastline, especially sheltered valleys, can be filled with the scent of balsam bog.

Wild Celery

Apium australe

NATIVE

This plant is closely related to cultivated celery. It is 15–80 cm in height, depending upon the shelter and moisture of the location. The stems are ridged and have a typical celery flavour. The leaves grow off branched stems and are made up of a number of leaflets that are sharply indented with jagged edges. They vary in colour through all shades of green. Flowers appear above the plant in a white floret cluster with a flattened top (INSET). The seeds are light brown.

IN FLOWER: November to March.

PLACES TO LOOK: On coastal areas along the high tide mark and amongst tussac grass.

LOCAL INFORMATION: The stems and leaves can be used to flavour soups and stews where the often strong, bitter flavour of the raw plant becomes more subtle. (see recipe on page 83)

Crowberry family

Diddle-dee

Empetrum rubrum

NATIVE

Diddle-dee is a very common, low-growing, evergreen shrub that covers vast areas of the Falkland Islands. Individual plants are generally 10–50 cm in height and can spread over large areas. The tiny leaves crowd around the stem creating a bushy appearance. The stems and roots are woody and large root twigs can often be found amongst the vegetation. Male and female flowers are on different plants: male flowers are deep red with obvious red filaments (INSET *top*): female flowers are chestnut brown. The berries are round and range from deep, crimson-red to bright orange-red (INSET *bottom*).

IN FLOWER: October, the berries begin to ripen in January.

PLACES TO LOOK: Very common throughout the Falkland Islands particularly on dry peaty soils.

LOCAL INFORMATION: The berries are a valuable winter food source for birds including the upland goose and Falkland thrush. They have a sour flavour but are made into jam and jelly, and also used to flavour yoghurt and cheese. (see recipes on pages 88, 89 and 91)

Daisy family

Buttonweed

Leptinella scariosa

NATIVE

A creeping plant, no more than 2 cm in height that can cover vast areas of ground with its long, straggly stems up to 12 cm in length. The bright green leaves are feathery and fern-like with small leaflets growing on either side of the central stem. Pale yellow flowers are small, round and compact although appear almost fluffy at the top. The seed capsules are unusual: pale green and rounded, quite tall compared to the rest of the plant. They have a distinctive pattern on the top.

IN FLOWER: December to January.

PLACES TO LOOK: Damp, coastal greens and alongside ponds and streams.

LOCAL INFORMATION: It is possible that this plant got its name from the rounded seed capsules that look rather like buttons. In gardens it is an invasive weed and is almost impossible to clear due to its extended, spreading root system.

Christmas Bush

Baccharis magellanica

NATIVE

A low-growing, spreading evergreen shrub with separate male and female plants. It grows to around 8 cm in height and can reach over 1 m in diameter. The stems are rough and woody with many branches spreading across the ground. The leaves, in a rosette formation around the stem, are light-green, rounded and quite succulent. Masses of delicate, creamy-white flowers appear above the leaves at the end of the stems. Cream coloured, fluffy parachute seed heads form on the female plants (INSET) and are dispersed by the wind.

IN FLOWER: December to January.

PLACES TO LOOK: Coastal heathland, particularly amongst diddle-dee. Newly disturbed dry, stony ground.

LOCAL INFORMATION: This is called the Christmas bush because its flowers appear in profusion around Christmas time, creating vast mats of creamy white flowers. It is locally used in the same way that mistletoe is in the Northern Hemisphere for those festive romantic moments!

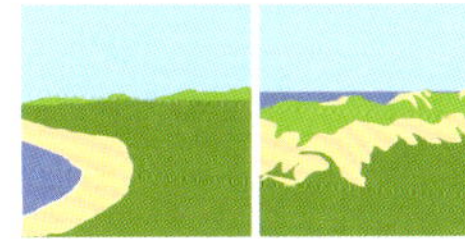

Sea Cabbage

Senecio candidans

NATIVE

This is a striking, silver leaved plant that colonises the edges of sand beaches. It can reach up to 1 m in height in more sheltered locations such as sand dunes. The leaves are large, fleshy and oval-shaped with crinkled edges and a soft, woolly, silvery surface, with dense fine white hairs. The compact, rounded, bright yellow flowers (INSET) appear in clumps at the top of a broad stem. The seed heads are round and brown.

IN FLOWER: December to February.

PLACES TO LOOK: Found on nearly every sand beach in the Falkland Islands, just above the high tide mark, and on coastal greens.

LOCAL INFORMATION: This plant was eaten by early sailors due to its anti-scorbutic properties which helped prevent scurvy, a disease caused by lack of vitamin C. It certainly doesn't taste as good as a home-grown garden cabbage!

Daisy family

Marsh Daisy
Aster vahlii

NATIVE

This plant generally grows in obvious clumps in damp peaty areas. The flowering stems are quite broad and can be up to 30 cm in height. They vary in colour between dark green and red-brown. The leaves, which are long, narrow and dark green, grow along the length of the stem becoming smaller at the top. The flower buds are pink-purple which adds to the overall colour of the plant. On opening, white daisy-like flowers appear with many petals around a yellow centre. The seeds are white feather parachutes, dispersed by the wind.

IN FLOWER: January and February.

PLACES TO LOOK: In damp white grass areas and shrub heathland.

LOCAL INFORMATION: It is easy at first glance to mistake this for the introduced species of European daisy but the clumps it forms are much taller and appear almost bushy with more prominent leaves.

Vanilla Daisy

Leucheria suaveolens

ENDEMIC

This is a striking plant with beautiful, creamy white flowers at the top of a single stem, which can grow up to 25 cm. The leaves are pale green-grey with a covering of fine hairs and are deeply indented with individual leaflets along the edge of each leaf. They form a rosette-shaped cluster at the base of a broad woolly stem. The flowers have numerous slender, white petals around a tight, pale golden centre and have a distinctive vanilla scent, which is particularly fragant on a warm summer's day. The seeds (INSET *bottom*) appear as rounded clusters of white parachute seed heads.

IN FLOWER: November to January.

PLACES TO LOOK: In coastal heathland amongst white grass, diddle-dee and fern beds.

LOCAL INFORMATION: The strong and distinctive vanilla scent makes this a popular flower with everyone especially, children. They say it is just like ice cream!

Fachine

Chiliotrichum diffusum

NATIVE

This large evergreen bush is one of the biggest native plants to be found in the Islands. It can grow to 2 m in height in sheltered valleys, and can spread to 2 m. The small, pointed oval leaves are grey-green and coated with white hairs, particularly on the underside. The upperside is a darker shade of green. The stems become quite woody with age. The flowers are like large daisies with white petals around a yellow centre (INSET). The seeds are delicate white feathered parachutes dispersed by the wind. The plant has a distinctive aromatic herby smell.

IN FLOWER: December to February.

PLACES TO LOOK: Sheltered valleys and damp heathland areas.

LOCAL INFORMATION: The presence of fachine is an indicator of thriving native vegetation as it does not tolerate grazing by sheep and cattle. The aromatic leaves can be used to flavour soups and stews. (see recipe on page 85)

Daisy family

Groundsel

Senecio vulgaris

INTRODUCED

This medium-sized, bushy plant can grow. up to 40 cm in height. Pale green leaves are quite fleshy and have deep indentations giving a distinctive feathery, almost fern-like appearance to the foliage. Small, yellow, cylindrical shaped flowers appear in clusters at the top of the stem (INSET). White parachute seeds are dispersed by the wind.

IN FLOWER: September to May.

PLACES TO LOOK: Widespread in dry locations and often a primary coloniser of eroded or burnt areas.

Sticky Groundsel *Senecio viscosus*

This closely-related species, also introduced from Europe, is found in similar places, particularly around Magellanic penguin burrows. It differs from groundsel in being wholly covered in sticky hairs and having a aromatic, lemon scent when crushed. The similar yellow flowers lack the black-tipped bracts of groundsel.

LOCAL INFORMATION: This plant was introduced from Europe and is now widespread throughout the Islands. It has value as it stabilises bare soil in areas which have been burnt or overgrazed. However this could prevent native species from re-colonising such areas.

Antarctic Hawkweed

Hieracium antarcticum

NATIVE

A relatively tall flowering plant with yellow flowers. The slender flowering stems, sometimes a reddish colour, can reach 30–40 cm in height. The leaves are long, broad and pointed with jagged indentations along each edge, with a very obvious covering of fine hairs. They form a flattened cluster at the base of the plant. The single yellow flowers (INSET) have numerous narrow petals. White parachute seeds are dispersed by the wind.

IN FLOWER: December–February.

PLACES TO LOOK: Coastal areas and along roadside verges in Stanley.

LOCAL INFORMATION: This plant is like the dandelion in many ways. It colonises grassland areas and is an annoying weed in many gardens, although it is easier to remove than the dandelion.

Daisy

Orang

Hierac

A pla
flower
flower
in hei
confir
a cove
are alı
dark l
at the
have
shape

LOC
spla

ERRATUM

Plants of the Falkland Islands by Ali Liddle

The photographs and text on page 40 refer to the introduced species Common Cat's Ear *Hypochaeris radicata* and not to the native Antarctic Hawkweed *Hieracium antarcticum* as stated. We apologise for this error.

INTRODUCED

ıute seeds are dispersed by the

ɪecember to March.

OK: Commonly found
ssy roadsides in Stanley, but
; to other areas of coastal
d white grass nearby.

weed, it makes an attractive

Daisy family

Falkland Lavender

Perezia recurvata

NATIVE

A very attractive low-growing, evergreen shrub that forms large, spreading clumps. It may reach 30 cm in height in sheltered locations and have a spread of up to 1 m. The leaves are dark green, narrow and spiky and tend to cluster at the top of the woody stems. These feel quite hard and prickly to the touch. The slender, curled petals, range from deep lilac to pale blue and even white. The seeds (INSET) are fluffy, cream-coloured parachutes dispersed by the wind.

IN FLOWER: December to January.

PLACES TO LOOK: Dry coastal areas and heathland.

LOCAL INFORMATION: Falkland lavender is not related to the lavenders of the Mediterranean and has no distinctive lavender scent. It was apparently named because of its colour alone.

Daisy family

Coastal Nassauvia

Nassauvia guadichaudii

ENDEMIC

An attractive plant which grows in dense, low-growing cushions up to 10 cm in height with a spread of up to 1 m. The dark green, rosette-shaped leaves have spines along their edges and cover the surface of the cushion. Vegetation from previous years' growth forms the base of the plant. The flowers, are small and cream with five petals forming a daisy-like shape and appear on the surface of the cushion. They have a strong scent of nectar which is very noticable on a warm summer's day.

IN FLOWER: December to February.

PLACES TO LOOK: Dry, rocky coastal areas.

LOCAL INFORMATION: This endemic is widespread and common in the Falkland Islands.

Falkland Smooth Ragwort

Senecio vaginatus

ENDEMIC

Ragworts
The two endemic Falkland ragworts should not be confused with two introduced species. Rare in the Falklands, European Ragwort *Senecio jacobaea* and Oxford Ragwort *S. squalidus* are both taller plants with very indented leaves that grow in rosettes in their early stages. European Ragwort is highly toxic to livestock and should be eradicated when found. Do take care to make absolutely sure it is not one of the endemic ragworts (non-toxic) before removal. For more information refer to the Falklands Conservation leaflet *Invasive Plants Management and Removal.*

This plant has distinctive yellow, daisy type flowers. The single flower stems reach 20 cm in height and often branch at the top to produce two or three flower heads. The leaves, which appear along the length of the stem, are small, narrow and pointed. Leaves and stems are both yellow-green in colour, and are quite succulent. Some of the upper leaves have delicate covering of fine hairs. The flowers have many slender, oval, overlapping petals around a deeper yellow centre. Seeds are creamy coloured parachutes, dispersed by the wind.

IN FLOWER: November to February.

PLACES TO LOOK: Diddle-dee heath, fern beds and dry rocky areas.

LOCAL INFORMATION: This plant is one of the 13 endemic plants in the Falkland Islands, the only place in the world where they grow. It may be confused with the more common Falkland woolly ragwort which grows in clumps, but has more obvious white hairy leaves and fewer petals.

Daisy family

Falkland Woolly Ragwort

Senecio littoralis

ENDEMIC

This is a bushy plant with yellow daisy-like flowers. The flowering stems can reach 25 cm in height. Both the long, narrow, almost oblong pale green leaves and woody stems are coated with white hairs making them appear woolly. The flowers have slender yellow petals sparsely arranged around a deeper yellow centre. They appear on numerous stems above the leaves creating a bushy cushion of colour. The seeds appear in delicate clusters of white fluffy parachutes (INSET).

IN FLOWER: November to January.

PLACES TO LOOK: Both coastal and inland areas amongst diddle-dee heath and fern beds.

LOCAL INFORMATION: The Falkland woolly ragwort can only be found in the Falkland Islands. It is distinguished from the Falkland smooth ragwort by its compact bushy formation, the sparse arrangement of petals on the flowers and the silvery-grey leaves. It brightens many coastal areas with large patches of yellow across the landscape.

Dock family

Sheep's Sorrel

Rumex acetosella

INTRODUCED

This is a very common plant which colonises large areas of ground. Its height depends on location but ranges between 5–20 cm. Leaves are bright green, slender and almost arrow-shaped and are confined to the base of the plant. Red flowers appear as little bunches of rounded florets towards the top of stem and can make spectacular splashes of colour throughout camp in autumn.

IN FLOWER: November to February.

PLACES TO LOOK: Open sand and shingle areas near the high tide mark, eroded peat areas and heathland. It is particularly noticeable on overgrazed tussac islands.

LOCAL INFORMATION: Despite being an introduced species, sorrel is useful as a primary coloniser of areas damaged by overgrazing. It stabilises the soil allowing further colonisation by native species. The leaves are edible and can add a lovely flavour to salads. (see recipe on page 84)

Figwort family

Boxwood

Hebe elliptica

NATIVE

This evergreen shrub is one of the tallest native plants in the Islands. It grows up to 2·5 m in height with a spread of over 2 m at times. The leaves are oval-shaped and shiny and are neatly arranged in groups of four around a woody stem. The flowers are white or, less commonly, lilac and crowd around the top of the stems. Each flower has four quite robust petals around very obvious stigma and stamen. The seed capsules and seeds are brown.

IN FLOWER: December to February.

PLACES TO LOOK: Inaccessible coastal cliffs and slopes and in many gardens in Stanley and settlements.

LOCAL INFORMATION: Boxwood used to be common along the coast of West Falkland, but, being susceptible to overgrazing by farm stock, is now only found in those coastal areas that are inaccessible to sheep. However, it is commonly used as hedging in gardens, although some people dislike it as 'blue buzzer' flies tend to shelter and breed in it.

Figwort family

Lady's Slipper

Calceolaria fothergillii

ENDEMIC

A small perennial plant with distinctive orange and yellow 'slipper'-shaped flowers. The short, red stem is covered with fine white hairs and is around 12 cm in height. The oval leaves are covered with fine hairs. The orange flowers have a deep, red stripe around the edge of the larger lower lip while the smaller upper petals are marked with red spots and streaks. The seed capsule is brown and cylindrical.

IN FLOWER: November to December.

PLACES TO LOOK: Widespread and locally fairly common among diddle-dee heath and on open well drained sandy soils or where peat cover is thin. Can also be found on coastal slopes.

Yellow Lady's Slipper *Calceolaria biflora*
This similar species (BELOW), occurs rarely and is protected by law. It differs from lady's slipper in having 2–4 flowers on each stem and the lower lip of each flower lacks the distinctive red stripe of lady's slipper.

LOCAL INFORMATION: It was only in 2004 that this plant was discovered to be endemic to the Falkland Islands.

Felwort

Gentianella magellanica

NATIVE

Felwort is a relatively small plant that is quite bushy when in flower. The broad flowering stems reach around 12 cm in height and are a distinctive dark purple. The leaves are a slender, triangular shape and bright green. Some purple shading may be seen along the edges, particularly towards the tip. The lilac flowers appear in clusters at the top of the stems. Each of the five petals is marked with deeper purple streaks leading to a green centre.

IN FLOWER: January and February.

PLACES TO LOOK: Found in low-growing shrubby heathland, acid grassland and rocky areas.

LOCAL INFORMATION: This is a beautiful little flowering plant that is fairly common but often overlooked amongst taller vegetation. There are thriving populations above Stanley racecourse and on the edge of Yorke Bay pond near Stanley airport.

Giant-rhubarb family

Pig Vine

Gunnera magellanica

NATIVE

A common, low-growing, ground cover plant found throughout the Falklands. It is 5–20 cm in height, depending on location. Distinctive kidney-shaped leaves are shiny and wrinkled with deep veins. The undersides of the leaves are hairy. Female flowers (INSET *top*) are deep pink, rounded fluffy clusters found near the base of the plant. Male flowers (INSET *middle*) are tall, red spikes with sparse flowers along the length of the stem. The berries (INSET *bottom*) are brilliant red-orange and found in clusters at the top of a short stalk.

IN FLOWER: October to December. Berries ripen February to April.

PLACES TO LOOK: Abundant, but particularly common in sheltered, coastal areas with running water and in damp white grass.

LOCAL INFORMATION: The berries are poisonous. In sheltered areas pig vine can appear quite large and bushy but amongst short grass it grows very flat to the ground.

Mountain Berry
Gaultheria pumila

NATIVE

This is a small, low-growing, evergreen shrub. It is a creeping plant, no more than 4 cm in height but its stems can be up to 50 cm in length. The small oval-shaped leaves are green and shiny, and grow in an opposite formation around the stem. The flowers are white, bell-shaped, with a distinctive red sepal on stems that curl above the leaves. The berries (INSET) are large, round, deep pink and often lie very close to the ground.

IN FLOWER: November to December. The berries ripen in March.

PLACES TO LOOK: It is commonly found throughout the Falkland Islands but prefers sandy coastal slopes, mountain ridges and heathland.

LOCAL INFORMATION: Don't confuse this with teaberry – the berries are not poisonous but they really don't taste half as good! It is often found growing through balsam bog plants.

Pale Maiden

Olsynium filifolium

NATIVE

An elegant, slender plant with lovely, white, bell-shaped flowers. The long, thin flowering stem is around 20 cm in height. Grass-like leaves curl upwards around the lower part of the stem. The beautifully scented flowers which appear at the top of a single stem have six petals, marked with pale purple stripes leading to a yellow centre. Seed capsules are round, brown and explode scattering seeds over a wide area.

IN FLOWER: November to February.

PLACES TO LOOK: White grass and coastal heathland areas.

LOCAL INFORMATION: This Falkland plant is a race locally distinct from similar South American species. It is the National Flower of the Falkland Islands. Beautiful carpets of pale maidens still appear in a few coastal areas in springtime. Yellow maiden (page 53) is a closely related plant but is very rare in comparison.

Yellow Pale Maiden

Sisyrinchium chilense

NATIVE

Yellow pale maiden is closely related to the pale maiden but is much rarer. It is a delicate plant with a single flower at the top of a relatively broad stem which can reach around 15 cms in height. The stem, which broadens near the top, appears rather flattened and has obvious reddish markings along its length. The delicate yellow flower (INSET) has six slender petals which narrow to a point at the ends. The underside of each petal is marked with brown lines. Seeds are contained in a rounded seed capsule.

IN FLOWER: December to February.

PLACES TO LOOK: White grass, coastal greens, heathland.

LOCAL INFORMATION: Only three populations are known on East Falkland where it is thought to have been introduced. It has a much more widespread distribution on West Falkland, but is not common and is a plant protected by law.

Myrtle family

Teaberry

Myrteola nummularia

NATIVE

Teaberry is a low-growing shrub. It rarely reaches more than 3 cm in height but the creeping stems may be up to 15 cm long. Small, rounded, shiny leaves are dark green and almost waxy in appearance. They are found along the length of the stems. The flowers are small and white with four rounded petals around a pale pink centre. Teaberries ripen to form large pink and white juicy berries (INSET).

IN FLOWER: November to January.
The berries ripen in March and April.

PLACES TO LOOK: Found in damp, peaty areas amongst white grass and astelia.

LOCAL INFORMATION: This plant gets its name from the early settlers who used the leaves to make tea. The berries have a sweet flavour all of their own and are a traditional treat although being low-growing, often amongst prickly astelia, make them a well earned one. They are delicious with fresh cream, in buns and pavlovas. There are well guarded recipes in a number of households for teaberry vodka! (see recipes on pages 86, 87 and 91)

Orchid family

Dog Orchid

Codonorchis lessonii

NATIVE

This plant has a distinctive white flower on a tall single stem. The reddish stem is 10–25 cm in height. Three rounded, shiny green leaves form a cluster at the base of the stem. The flower (INSET) has three slender, white petals around a central area and labellum (lower lip), which are marked with purple dots. It has a pleasant scent. The seed capsules are brown and cylindrical in shape.

IN FLOWER: November to December.

PLACES TO LOOK: White grass camp and diddle-dee heathland.

LOCAL INFORMATION: This beautiful flower is the most common orchid in the Falkland Islands but is sometimes difficult to find as it only flowers for a short time.

Gaudichaud's Orchid
Chloraea gaudichaudii

NATIVE

This plant produces an unusual pale green, almost white flower. The stem can grow to 25 cm in height. Its leaves are dark green with obvious veins, and spiral round a rather flattened stem. Two or three flowers appear at the top of this broad stem. Triangular outer petals are pale green and have very definite dark green veins while the central petals are pale yellow with green pimples along their length. The cylindrical seed capsules can contain thousands of tiny brown seeds.

IN FLOWER: November to December.

PLACES TO LOOK: Generally found amongst damp areas of diddle-dee heathland and white grass.

LOCAL INFORMATION: This plant can be difficult to find and identify. The flower may not open, particularly in dry weather, but the plant can self-fertilise without opening its flowers.

Orchid family

Pale Yellow Orchid

Gavilea australis

NATIVE

This distinctive orchid grows up to 30 cm in height. It has long pointed leaves, which vary in colour from grey-green to bright yellow-green. These circle around the main stem making it appear quite broad. Beautiful creamy-white flowers cluster around the top of the stem and open in sequence from the bottom. The petals are marked with green veins, streaks and spots and surround a pale yellow centre. The seed capsules have slender oval-shaped pods and explode to scatter a vast number of tiny brown seeds.

IN FLOWER: December to January.

PLACES TO LOOK: Damp sandy coastal areas and diddle-dee heathland.

LOCAL INFORMATION: This orchid is rare and protected by law. It is illegal to dig up, pick or harm the plant in any way. In recent years it has flourished in an area near Stanley airport where it now covers quite extensive areas.

Orchid family

Yellow Orchid

Gavilea littoralis

NATIVE

This is the tallest orchid to be found in the Islands reaching up to 40 cm in height. The dark green leaves are a slender oval shape and spiral upwards around the lower part of the stem. Its very striking, deep yellow flowers have delicate green lines along each petal while the larger central petal (labellum) has an orange stripe down the centre with more obvious green markings. Flowers are set at regular intervals up the main stem. The plant produces brown, oval seed capsules.

IN FLOWER: November to January.

PLACES TO LOOK: Dry clay areas on coastal cliffs and some heathland areas, often growing in loose colonies

LOCAL INFORMATION: This is an uncommon plant that can be difficult to find and identify as the flowers often fail to open. When this occurs the plant is capable of self-fertilisation.

Pea family

Gorse

Ulex europaeus

INTRODUCED

This evergreen shrub, up to 2 m tall and with a spread of up to 10 m, has a covering of bright yellow flowers in early spring and summer. The lower, older stems appear as thick woody branches while the upper stems are covered in thin, sharp thorns. These upper stems are densely covered in yellow 'beak-shaped' flowers (INSET) which have a distinctive sweet scent (a mixture of honey and coconut). The small oval-shaped leaves are dark green but are not easy to see during the flowering period. Seeds develop in small 'pea pods' which explode when dry, scattering the seeds over a wide area

IN FLOWER: October to January.

PLACES TO LOOK: Around settlements and farms, and in Stanley.

LOCAL INFORMATION: This plant was introduced by European settlers to create hedges to contain cattle. It is found in most settlements and provides shelter for animal stock. Its tendency to spread can cause problems as it can encroach on grazing pastures. The native green spider can often be found on gorse, while meadowlarks, thrushes and barn owls will nest within its dense cover. (For further information see Falkland Conservation's leaflet *Invasive Plants Management and Removal*)

Philesia family

Almond Flower

Luzuriaga marginata

NATIVE

This is a low-growing, creeping plant with flattish, almond-shaped leaves which grow in an opposite formation on the sides of the stem. It can reach 4 cm in height but the stems can trail for up to one metre along the ground. White flowers have six oval petals and a yellow stigma. The seeds are contained in round, dark red fruit. (INSET)

IN FLOWER: November to January.

PLACES TO LOOK: Particularly favours the edge of stone runs and flourishes in shaded sites. It is also found on mountain slopes, coastal cliffs and amongst fern beds.

LOCAL INFORMATION: The plant is named because of the strong and distinctive almond scent of the flowers, but the leaves also look like little almonds as well.

Pink family

Field Mouse-ear

Cerastium arvense

NATIVE

This appears as a straggling plant although in favourable conditions it can grow in quite dense clumps creating small patches of colour in camp. It is usually less than 5 cm in height but can spread up to 25 cm. The small oval leaves are often sparse and yellowing particularly on the lower part of the stem. The flowers are white and have five petals but because they have a deep indentation along the outer edge at first glance they can appear to have more. Each flower has a pale yellow centre.

IN FLOWER: November to February.

PLACES TO LOOK: Coastal heathland and white grass.

LOCAL INFORMATION: This plant is often overlooked where it grows amongst white grass and diddle-dee. In spring and summer its white flowers are easily seen.

Primrose family

Dusty Miller
Primula magellanica

NATIVE

This plant produces beautiful white flowers on a stem 10–20 cm in height. The grey-green curved leaves cluster around the base of a single, thick stem. Both the stem and leaves have a covering of soft white hairs which gives them their grey colouring. The flower is a rounded white globe made up of a cluster of individual primrose flowers. Each floret has four heart-shaped petals around a yellow centre. The tiny dust-like seeds are contained in a round capsule that dries and splits, scattering seeds over a wide area.

IN FLOWER: October to December.

PLACES TO LOOK: Dry, sandy, open heathland.

LOCAL INFORMATION: This beautiful white flower was more common but has suffered, as many plants have, from overgrazing. It is the only primula to grow wild in the Falkland Islands.

Primrose family

Pimpernel

Anagallis alternifolia

NATIVE

A very low-growing, branching plant which colonises large areas of damp, sandy coastal ground. As a ground cover plant it is no more than 1–2 cm in height. The oval-shaped, bright green leaves are quite fleshy in texture and appear along the length of the stems. Distinctive pale pink, star-shaped flowers have a dark pink centre with delicate pink stripes along each petal. Each flower has five petals around a yellow central stamen.

IN FLOWER: December to February.

PLACES TO LOOK: Damp sandy areas above the high tide mark and on coastal slopes and cliffs.

LOCAL INFORMATION: Pimpernel forms extensive pink carpets along the coastline throughout the summer, particularly in January.

Rose family

Falkland Strawberry

Rubus geoides

NATIVE

A very low lying, ground cover plant which is no more than 2 cm in height. The small, shiny oval-shaped leaves have very obvious toothed edges. Conspicuous white flowers have six rounded petals and a yellow-green centre. Both the leaves and flowers are similar to that of a cultivated strawberry although much smaller and shiny green beneath. The distinctive red berries (INSET) are made up of shiny clusters of seeds and look more like raspberries than strawberries. These berries tend to nestle very close to the ground, tucked in amongst the leaves. They are pushed into the leaf litter by their stalks

IN FLOWER: November to December.
The berries appear in January and February.

PLACES TO LOOK: White grass and heathland areas, amongst rocky outcrops and often between tussac grass bogs.

LOCAL INFORMATION: The berries are delicious: a combination of raspberry and strawberry. They are best served simply with fresh cream. The location of patches of native strawberries are often closely guarded, family secrets!

Rose family

Prickly Burr

Acaena magellanica

NATIVE

This plant grows in loose clumps with long straggling stems. It is 20–40 cm in height, depending on the location. Both stems and leaf stalks are red and each leaf is made up of 4–7 pairs of leaflets with serrated edges. Multiple flowers appear as reddish green globes at the top of a tall thick stem. The distinctive seed burrs are made up of many individual seedcases with harpoon-headed spines (INSET) which attach themselves to passing birds and animals.

IN FLOWER: November to February.

PLACES TO LOOK: On coastal greens and on white grass areas near ponds and streams.

LOCAL INFORMATION: This plant can cause problems for sheep farmers because the burrs attach themselves to a sheep's fleece which reduces the market value of the wool. The shearers don't like them much either because of the prickles! Children in the Falklands call them 'Sticky Pom Poms' which is a very accurate description.

Stonecrop family

Stonecrop

Crassula moschata

NATIVE

A small, creeping succulent plant that covers large areas of sandy coastal greens. As a ground cover plant it is no more than 2 cm in height but can creep over 1 m. The leaves are small, oval-shaped and grow in opposite formation on long trailing stems. The leaves vary significantly in colour from deep red to light green. Flowers (INSET) are tiny, either white or pink and appear near the top of the stem.

IN FLOWER: December to February.

PLACES TO LOOK: On sandy and rocky shores just above the high tide mark. It can also be found trailing down damp rock faces.

LOCAL INFORMATION: This plant is unusual as it can appear in two very different forms, which can be confusing. It appears as a red ground cover plant, quite sparse in nature, and also as a lush, green plant trailing down damp rock faces near the shore.

Sundew family

Sundew
Drosera uniflora

NATIVE

This plant lies flat to the ground and only reaches 3 cm in height. It produces white flowers but it is the sticky globular red hairs on the small round leaves that are its most noticeable feature and give the plant its red appearance. The sticky hairs trap small insects, the plant releases a substance that digests the prey and the resulting 'soup' is absorbed into the plant. The flowers are quite large in relation to the size of the plant with five white petals and a yellow-green centre. They grow on a single stem and appear above the glistening leaves.

IN FLOWER: January to February.

PLACES TO LOOK: Wet, peaty areas beside ponds and streams.

The photo shows sundew growing amongst astelia, a common wetland plant with robust, pointed leaves.

LOCAL INFORMATION: This is the only insectivorous plant in the Falkland Islands. In summer the leaves glisten in the sunshine, hence the name sundew.

Thrift family

Falkland Thrift

Armeria macloviana

NATIVE

This is a maritime plant which grows in dense stocky clumps. These vary in size, as do the flowering stems, which can be up to 15 cm in height, depending on the location. The leaves are dark green, long and narrow, rather like a grass. The round, rose pink flowers appear at the top of tall single stems. These dry to become a rusty brown colour through the summer (INSET). The seeds form at the petal base and are dispersed by the wind.

IN FLOWER: November to January.

PLACES TO LOOK: Found in most coastal areas above the high tide mark and on cliff edges where the soil is dry and sandy.

LOCAL INFORMATION: This is a very common plant around the coast of the Falkland Islands and can even be found along the sea wall in Stanley.

Common Violet

Viola maculata

NATIVE

This is a delicate plant with bright yellow flowers. The leaves are oval, shiny and with gently toothed edges. Three or four leaves appear in a cluster around the base of the flower stem. The flowers have five oval petals and are held on a single stem up to 15 cm in height. The larger, lower petal is streaked with brown markings, which lead to the centre of the flower. The seed pods are cylindrical and brown and, when dry, explode to disperse the seeds.

IN FLOWER: November to January.

PLACES TO LOOK: Dry coastal heathland particularly on dry peat areas or dry sandy slopes.

LOCAL INFORMATION: This violet is thought to be the main food plant for caterpillars of the Queen of the Falklands fritillary, the only resident butterfly in the Falkland Islands.

Water-milfoil (Pondweed)
Myriophyllum quitense

NATIVE

This freshwater pond weed often completely covers the surface of Falkland ponds forming dense red carpets. The stems are as long as the depth of the water with most leaves appearing on the upper part of the plant. Below the surface of the water the leaves are green and feathery, whilst at the surface they are oval and a deep rusty red. The solitary purplish-green flowers are very tiny (about 2·5 mm long).

IN FLOWER: January to February.

PLACES TO LOOK: Fresh water ponds and slow-flowing streams.

LOCAL INFORMATION: Water-milfoil provides valuable food for black-necked swans and other waterfowl. The aquatic invertbebrates that live amongst the weed, such as water beetles and freshwater snails, are important in the diet of many other waterbirds.

Scurvy Grass

Oxalis enneaphylla

NATIVE

This is a common plant found throughout the Falkland Islands. Its height varies according to location but can be up to 20 cm if sheltered by other plants such as diddle-dee. The leaves are light grey-green and radiate from a central point to create a distinctive leaf cluster. The pale red stems are slender and can be long and straggly. Flowers range from white to pale pink to deep mauve (INSET), with white being the most common colour. Each flower has five curved petals radiating from a yellow centre.

IN FLOWER: October to February.

PLACES TO LOOK: Coastal heathland and amongst diddle-dee and white grass.

LOCAL INFORMATION: This plant is edible. Both the leaves and flowers have a pleasantly sharp, citrus flavour and have antiscorbutic properties. It was eaten by sailors in the past to prevent scurvy. Refreshing drinks can made from this plant and it adds a wonderful tangy flavour to salads. (see recipes on pages 84 and 90)

Cypress family

Monterey Cypress

Cupressus macrocarpa

INTRODUCED

Trees

The largest forest in the Falkland Islands can be found at Hill Cove on West Falkland. It is owned and protected by the Falkland Islands Government.

Planting started here in the 1800's but many of the trees seen today, mainly pines, spruces, poplars and Patagonian beeches, were planted in the 1920's. This plantation is proof that not all the trees in the islands are stunted and windblown, some here have now grown to 15 metres tall.

In recent years, however many of the trees have begun to die off. There is evidence of spruce aphids, which slowly kill trees but it could also be that they are coming to the end of their natural life. There has been little natural regeneration so the days of a flourishing forest at Hill Cove could be numbered.

This is a medium-sized evergreen tree, which often becomes an irregular flat-topped shape due to the wind. It grows to heights of 10–20 m with a trunk diameter up to 6 m. The foliage grows in bright green dense sprays. The leaves are quite scaly and appear on long rounded shoots around 20 cm in length. The seed cones (INSET) are hard and round and made up of roughly hexagonal platelets that open when dry to release the seeds. They are found in clusters towards the end of the branches. This tree was introduced from the coast of California, an area of strong, drying winds, which is why it has adapted so well to the Falklands' climate.

IN FLOWER: The cones are green at first, maturing to brown two years after pollination.

PLACES TO LOOK: Stanley, particularly in the old part of town, and camp settlements.

LOCAL INFORMATION: There are many mature trees to be found in Stanley. A number of impressive specimens can be found in the grounds of Christchurch Cathedral and Government House. Some settlements in camp boast mini 'forests' where trees have been planted as shelterbelts.

Cinnamon Grass

Hierochloe redolens

NATIVE

This is a distinctive grass which forms very large, glossy clumps. The shiny leaves, which are quite broad for a grass can reach up to 60 cm in length and have a pungent spicy aroma when crushed. The flowers and seed heads form shiny clusters at the end of long stems.

IN FLOWER: September to January.

PLACES TO LOOK: Coastal slopes and damp heathland areas.

LOCAL INFORMATION: This grass is susceptible to over-grazing by sheep so is often only found on un-stocked areas of coastline. Mutton from sheep fed on this grass has a distinctive 'cinnamon' flavour. A ready seasoned roast!

Marram Grass

Ammophila arenaria

INTRODUCED

This is a tall grass with long, thin leaves. It can grow to over 1·5 m. In summer the leaves are shiny and green but through the winter these die off leaving brownish, straw-like fronds. The large seed heads appear at the end of long stiff stems above the leaves. It grows in large clumps colonising extensive areas of sand dunes.

IN FLOWER: December to February.

PLACES TO LOOK: Sand dunes.

LOCAL INFORMATION: This is known locally as sand grass as it colonises and stabilises sand dunes. It can provide valuable food and shelter for livestock, particularly cattle. In winter the seeds are favoured by many seed-eating birds including grass wrens and long-tailed meadowlarks.

Tussac Grass

Poa flabellata

NATIVE

Tussac grass can grow up to 3 m in height. Each plant is made up of masses of long, thin green leaves that can reach 2 m in length. These grow from a large, fibrous pedestal known as a bog. The pedestals are made up of dry decaying leaves from many years of growth. Some tussac plants may be over 200 years old. The large, green seed heads emerge in August and September. These dry during the summer months and the seeds are dispersed by the wind.

IN FLOWER: September to October.

PLACES TO LOOK: Ungrazed, protected coastal areas and offshore islands.

LOCAL INFORMATION: The name 'tussac' comes from the huge clumps or 'bogs' which are formed by this giant grass. There was once a fringe of tussac grass around most of the Falklands' coastline but due to fires and uncontrolled grazing less than 20% remains today, mostly on offshore islands. Its ability to remain green throughout the year led to overgrazing by sheep, cattle and horses. The Falkland Islands is one of the most important locations in the world for this plant. In a land without native trees, tussac provides a very important habitat for Falkland wildlife. Birds nest and feed here, sea lions find shelter, and many insects (some of them dependent on and unique to tussac) live in the grass. Falklands Conservation has a replanting programme to restore areas of tussac grass, particularly on the two main islands where it has been so depleted. (for further information see Falkland Conservation's leaflet *Tussac Grass Planting and Management*)

Grass family

White Grass
Cortaderia pilosa

NATIVE

This is a coarse, wiry, sandy-coloured grass. It grows to a maximum height of around 40 cm but this depends upon its location. In damp, fertile valleys it forms quite dense rounded tussocks. On extensive flat land, particularly in Lafonia (a large, damp, peaty area, of East Falkland) it has a looser formation.

IN FLOWER: December to January.

PLACES TO LOOK: Most areas of the Falkland Islands.

LOCAL INFORMATION: This is the most common grass found in the Falkand Islands. It gives a distinctive light-brown, sandy colour to the landscape. Its nutritional value is poor so farmers run relatively low numbers of sheep over vast areas of land in order to ensure sufficient feed. White grass provides shelter for ground-nesting birds, invertebrates and many smaller plants including more nutritional grasses.

Rush family

Native Woodrush

Luzula alopecurus

NATIVE

This is a small grass-like plant with a very distinctive flower head. The main stem is around 15 cm in height. Leaves are long, slender, pointed and vary in colour from red-brown to light green with small hairs growing along each edge. Pale brown, egg-shaped flowers are covered in fluffy white hairs, and can be found at the top of a central stem. Small brown seeds, contained within the rounded head of the plant, are slowly dispersed by wind.

IN FLOWER: November to January.

PLACES TO LOOK: Coastal heathland and inland rocky areas.

LOCAL INFORMATION: This plant is especially attractive on a sunny day when the hairs catch the light.

Short Rush

Rostkovia magellanica

NATIVE

This common rush is found on damp marshy ground, growing in loose clumps that can reach around 20cm in height. The leaves are long and slender, and grow in clusters made up of brown leaves from previous years growth and new green growth. The flowers are rather insignificant but this plant is easily recognisable from its seeds. A single, shiny dark-brown seed capsule (RIGHT) appears at the top of the stem.

IN FLOWER: October to November.

PLACES TO LOOK: Damp, marshy ground by ponds and on wetter white grass areas.

LOCAL INFORMATION: It is sometimes known as the brown rush. When driving across camp it is always worth avoiding areas of short rush as the ground here is likely to be soft and wet and may result in a serious 'bogging' of the vehicle.

Tall Rush

Marsippospermum grandiflorum

NATIVE

This plant is tall with slender, pointed leaves. It grows in loose clumps 35 cm in height. The leaves are tubular and tend to be dark green at the base, becoming reddish-brown towards the top. Dry brownish leaves from previous years' growth are found amongst the loose cluster of the plant. Flowers appear at the top of single stems and have two upwardly pointing crowns of petals.

IN FLOWER: October to November.

PLACES TO LOOK: Coastal heathland and damp, peaty white grass areas.

LOCAL INFORMATION: The presence of tall rush in camp indicates wet ground which should be avoided when driving as a 'bogging' may occur.

Hard Fern family

Tall Fern

Blechnum magellanicum

NATIVE

This tall fern is a coarse leaved plant with long leaf fronds. These can reach around 70 cm in height. The diameter of the top of the plant may be equally large. It forms big clumps in dense beds that cover extensive areas of ground. The young fern fronds (INSET) emerge as soft, pale green and bronze leaves that turn dark green and toughen throughout the summer. The whole fern clump is made up of new growth and dead fronds from previous years growth that remain at the base of the plant for a number of years.

IN SEASON: The ferns release ripe spores in the autumn months, March and April.

PLACES TO LOOK: Diddle-dee heath areas, in sheltered valleys and amongst rocky outcrops.

LOCAL INFORMATION: This plant is very common, apart from on Lafonia, far western islands and Cape Dolphin where it is probably too dry, particularly during the summer months. Smaller plants such as almond flower and scurvy grass grow in sheltered niches amongst the ferns.

Hard Fern family

Small Fern
Blechnum penna-marina

NATIVE

A low-growing fern that spreads to form dense mats across large areas of heathland. Although this is primarily a ground cover fern the leaves can vary in length from 5–15 cm depending on location. They are quite rigid with a deep central groove. There are fertile and non-fertile fronds: fertile fronds are larger. The young leaves (INSET) unfurl as reddish-brown and bright green fronds which turn dark green through the summer months. Spores appear on the underside of the leaves.

IN SEASON: The spores ripen in March.

PLACES TO LOOK: Coastal heathland amongst diddle-dee, white grass and tall fern beds.

LOCAL INFORMATION: In sheltered locations the small fern can produce quite long fronds but on open ground they are small and lie flat to the ground. This is one of the most common plants in the Falkland Islands.

Falkland Recipes

The following recipes are based on traditional Falkland cooking and all include a selection of ingredients from some of the native plants described in this book. None of these plants are rare or threatened but please remember to pick only the amount you require.

Vegetable Soup with Wild Celery

Sheep's Sorrel and Scurvy Grass Salad with Scurvy Grass Dressing

Mutton Casserole with Herby Fachine Dumplings

Teaberry Pavlova

Teaberry Buns

Diddle-dee Jam

Diddle-dee Jelly

Iced Scurvy Grass Tea

Scurvy Grass Cordial

Diddle-dee Gin

Teaberry Vodka

VEGETABLE SOUP WITH WILD CELERY

Soup is a wonderful thing. Almost any vegetables can be used and a wide variety of plants can be added to flavour the soup. My Mum's recipe for vegetable soup is simply " A bit of this and a bit of that". It always tastes fantastic, it is healthy and low fat too. Wild celery has quite a strong flavour when raw, compared to commercially grown versions, but when cooked it has a mild, more subtle flavour. Try the following recipe, you can alter the quantity of celery to suit your taste.

Ingredients

- 1 onion, chopped
- 2 carrots, peeled and diced
- 2 leeks, washed and sliced
- 2 medium potatoes, peeled and diced
- 15-20 stems and leaves of wild celery, washed and roughly chopped
- 1·5 litre (2¼ pints) vegetable stock. (vegetable stock cubes are the easy option!)
- A generous squirt of tomato puree
- A sprinkle of dried mixed herbs
- Salt and pepper

1. Fry the onion in a little oil and butter until soft.
2. Add the carrots, leeks and celery to the pan, cover and cook on a low heat for 10 minutes to gently 'sweat' the vegetables. Don't add the potato at this stage as it will stick to the bottom of the pan.
3. Add the tomato puree and other seasoning, pour on the stock and add the potatoes. Bring to the boil and simmer for 25 minutes or until the vegetables are tender. Check the flavour and add seasoning to taste.
4. This soup can be served as a 'chunky' version or you can liquidise it to make a smooth soup.
5. Sprinkle a few finely chopped celery leaves into the soup just before serving.

   You can add red lentils to this soup to make a more substantial meal but for 100g (4oz) of lentils you will need to add an extra 300ml (½ pint) of stock as they do absorb the liquid whilst cooking.

SHEEP'S SORREL AND SCURVY GRASS SALAD WITH SCURVY GRASS DRESSING

Sheep's sorrel has a distinctive flavour and can add a refreshing bite to a green salad. The 'citrus' taste of the scurvy grass will also liven up each mouthful. The quantities for the salad will obviously depend on how many people you are feeding. The following recipe will feed 4 as a very generous side salad.

Ingredients

SALAD:

- 1 small lettuce, roughly torn
- Quarter of a white cabbage, finely shredded
- A good handful of sheep's sorrel leaves
- 15-20 scurvy grass leaves
- Half a cucumber, finely sliced
- A handful of pumpkin seeds or sunflower seeds

DRESSING:

- 3 tablespoons olive oil
- 1-2 tablespoon white wine vinegar
- 15-20 scurvy grass leaves, finely chopped
- Salt and black pepper

1. If you make the dressing a few hours before it is needed the citrus flavour of the scurvy grass will develop a little more.
2. Combine all the salad ingredients adding the dressing just before serving. Toss gently.
3. Decorate the salad with scurvy grass flowers, which are also edible.

This salad goes particularly well with fish. Try the locally caught mullet or trout or some of the offshore fish such as toothfish or kingclip.

Sheep's sorrel and scurvy grass salad (above); mutton casserole with herby fachine dumplings (right); fachine in flower (left)

MUTTON CASSEROLE WITH HERBY FACHINE DUMPLINGS

With a history based on sheep farming mutton and lamb was the staple diet of the population of the Falkland Islands. Mutton was known locally as "365" because it was eaten every day of the year. In recent years as other meats and frozen products have become available it is eaten less often. Changes in lifestyle also mean there isn't always the time in the working day to cook large roasts and casseroles that require long, slow cooking.

Try this as a winter warmer at the weekend.

CASSEROLE

1. Heat a little oil in either a flame proof casserole dish or a large pan then gently brown the meat before adding the onion and carrots.
2. When the onion is soft add the stock, tomato puree, Worcestershire sauce and other seasoning.
3. Cover and simmer gently on a low heat for around 2 hours or place in the oven at 170°C (325°F) Gas Mark 3 for around the same time, stirring once or twice during cooking.

DUMPLINGS

While the casserole is cooking you can prepare the dumplings, which need to be added about 20 minutes before the casserole is ready.

1. Mix all the ingredients in a bowl and add enough cold water to make an elastic dough – around 4–5 tablespoons should be enough.
2. Divide the mixture into 12–14 portions and roll into small balls before adding them to the casserole.
3. Continue to simmer the casserole for a further 20 minutes.
4. When the casserole is cooked you may need to thicken the gravy a little (I usually cheat and mix a teaspoon of cornflour with a little cold water and add it to the gravy)

   Serve the casserole with potatoes and vegetables.

Ingredients

CASSEROLE:

- 450g (1 lb) diced mutton
- 1 onion, chopped
- 2 carrots, diced
- 2 bay leaves
- 1 tablespoon tomato puree
- A good shake of Worcestershire sauce
- 1 teaspoon of wholegrain mustard
- 600ml (1 pint) stock. Lamb stock cubes are readily available these days and do add authentic flavour.
- Salt and pepper

DUMPLINGS:

- 150g (6oz) self-raising flour
- 75g (3oz) suet
- 2-3 tablespoons of fachine*, finely chopped
- Salt and pepper

* When collecting the fachine, gather young shoots from the end of the stems as these will have more flavour. Strip the leaves from the woody stem and chop finely.

TEABERRY BUNS

Teaberries can be very tedious to pick as the plant is very low-growing and is often found amongst prickly astelia but they are well worth the effort. People living overseas often ask about the flavour of teaberries and the best I can come up with is a juicy combination of candy floss and cinnamon!! Try them and see for yourself.

Ingredients

100g (4oz) butter or margarine

100g (4oz) caster sugar

2 eggs, beaten

100g (4oz) self-raising flour

100g (4oz) teaberries *you can add more if you like for extra flavour.*

This recipe can be increased to 150g (6oz) and 3 eggs or doubled to 200g (8oz) and 4 eggs.

1. Place paper cake cases on a baking tray or in individual patty tins. Pre-heat the oven to 190°C (375°F) Gas Mark 5.
2. Cream the butter and sugar until the mixture is pale and fluffy.
3. Add the beaten eggs a little at a time, beating well after each addition.
4. Sieve the flour into the mixture and fold in using a metal spoon.
5. Carefully fold in the teaberries.
6. Spoon the mixture into the paper cases and place the tray on the middle shelf of the oven. Bake for 15–20 minutes until pale golden.

Eat and enjoy!

TEABERRY PAVLOVA

Homemade pavlovas are a bit of a treat as they do take a while to cook. They need a cool oven, allowing the pavlova to dry out rather than cook. In the past the bottom oven of the peat Rayburn stove was ideal. The mutton roast would cook in the main oven while the pavlova would cook in the cooler oven underneath.

Ingredients

- 4 egg whites (very fresh eggs make better pavlovas as the whites whisk up more easily)
- 200g (8oz) caster sugar
- 1 teaspoon white vinegar
- 1 level teaspoon cornflour
- 300ml (half a pint) double cream
- 400g (14oz) teaberries (or enough to fill the top of the pavlova)

1. Pre-heat the oven to 140°C, (275°F) Gas Mark 1.
2. Oil a sheet of greasproof paper and place it on a baking tray.
3. In a very clean, grease-free bowl whisk the egg whites until stiff and peaks remain standing.
4. Add the sugar a little at a time, ensuring the mixture is well beaten after each addition.
5. Gently fold in the vinegar and cornflour using a metal spoon.
6. Spread the meringue mixture in a circle on the greaseproof paper with the edges built up slightly to create a gentle hollow in the centre. Place in the oven for 2–3 hours until crisp and dry. Allow it to cool a little before carefully peeling it off the paper.
7. Whisk the double cream until stiff and spoon it into the centre of the meringue.
8. Pile the teaberries on top of the cream.

   The more the better!

DIDDLE-DEE JAM

Diddle-dee jam has a wonderful flavour but many people are often put off by the number of seeds in the jam and so prefer it as jelly. The berries can be a little sour and so the jam requires large quantities of sugar in order to both sweeten the flavour and allow the jam to set.

Ingredients

You need equal quantities of fruit and sugar. Alter the quantities accordingly. 1 kg = 2·2 lb (approx)

1·8 kg (4 lb) diddle-dee berries

1·8 kg (4 lb) sugar

1. Boil the berries in a large pan until tender. This will take around 30 minutes.
2. Use a hand held blender to liquidise the fruit or, alternatively squash the fruit with a potato masher before adding the sugar. This releases the juice and pectin.
3. Stir thoroughly then gently boil the mixture for around 2 hours, stirring regularly.
4. Test the jam by putting a small amount on a plate. Allow it to cool then gently push the surface. If it 'wrinkles' then it has reached setting point.
5. Pour the jam into clean, warm jars and seal.

This jam is delicious on bread, toast or scones. It is also wonderful with creamy natural yoghurt.

A real taste of the Falkland Islands.

Diddle-dee jam (left) and diddle-dee jelly (right)

DIDDLE-DEE JELLY

It is often thought that diddle-dee berries are at their best after 'Sports Week' in February when they are ripe and easy to pick. Ripe berries contain more juice and pectin which will ensure the jelly sets properly.*

Ingredients

Diddle-dee berries

Sugar

1. Place the berries in a large pan and add water so they are just about covered. Bring the fruit to the boil and simmer gently until the berries are soft and beginning break up.
2. Crush the fruit with a potato masher to ensure all the juice is extracted from the berries.
3. Allow to cool slightly then place the berries in a muslin jelly bag, a clean pillow case or even the leg of a clean pair of tights and hang over a bowl to allow the juice to drain out.
4. Leave overnight to drain, resisting the temptation to squeeze the bag as this will result in cloudy jelly.
5. Measure the collected juice into a saucepan and add 450g (1lb) sugar for each pint of juice.
6. Bring to the boil and simmer, stirring from time to time until the jelly begins to set. You can test for the setting point by placing a little of the jelly on a plate and allowing it to cool. Gently push the surface of the jam. If it 'wrinkles' then it is ready.
7. Place the jelly in clean, warm jars and cover.

This jelly can be difficult to set and won't ever be really firm. Adding a grated apple to the berries at the beginning may help or alternatively a commercially produced pectin could be used.

This jelly is perfect on toast and with scones for those people who don't like the seeds in diddle-dee jam, but it also makes a lovely accompaniment to roast mutton or lamb.

* Sports Week became a holiday tradition to to mark the end of the shearing season and give camp residents a chance to catch up and socialise at the end of a busy farming season. Camp Sports take place in settlements on both East and West Falkland each year and include horse racing, shearing competitions, dog trials and steer riding. There are dances each night and it is one of the social highlights of the year.

Drinks

As islands with a long history of sailing and sailors rum is often referred to as the local drink. If you are ever offered a 'hot one' or a 'cup of tea, che' you will find yourself sipping a very warming hot toddy with rum and sugar in varying amounts. The following recipes use some of the local plants and berries and vary in their alcoholic content!

Non-alcoholic

ICED SCURVY GRASS TEA

1. Add boiling water and sugar to the tea bag and make a good brew as you normally would.
2. Remove the tea bags and add a handful of scurvy grass to the tea.
3. Stir the mixture and leave to cool before chilling it in the fridge.
4. Before drinking, strain the tea to remove the scurvy grass. This is a refreshing drink on a hot summer's day.

Thanks to Nic Huin for this. Apparently whilst out on summer fieldwork for Falklands Conservation studying black-browed albatross he used to make this tea in the morning, place it in a bottle and hang the bottle in the sea to cool. It was a refreshing drink at the end of the day.

Ingredients

1 tea bag per person

Sugar to taste

A good handful of scurvy grass leaves, flowers and stalks.

SCURVY GRASS CORDIAL

1. Half fill a large pan with the scurvy grass and add the sugar.
2. Pour the boiling water over the leaves and stir to dissolve the sugar.
3. Leave overnight to cool and allow the flavour to develop.
4. Strain the liquid into a jug.
5. Adjust the sweetness and strength of the drink by adding sugar or water respectively.
6. Serve with ice and a slice of lemon.

Ingredients

Scurvy grass - leaves, flowers and stalks

2 litres boiling water

125g (4oz) sugar

Alcoholic

These drinks don't have exact quantities or fermenting times, so alcoholic content will vary from bottle to bottle and year to year. You've been warned!

DIDDLE-DEE GIN

You will need a spare, empty bottle to divide the gin between the two bottles.

1. Empty the gin into a jug.
2. Gently crush the diddle-dee berries to allow the juice to be released and then half fill each bottle with the crushed fruit. A bit of a tricky procedure!
3. Share the sugar between the two bottles and top up each bottle again with the gin.
4. Shake the bottles well and place in a dark cupboard for around 3 months.
5. Shake the bottles every now and again. The longer you leave this the stronger it will become both in flavour and alcoholic content!
6. When drinking, pour a measure of gin through a tea strainer and return the fruit to the bottle. You can drink this on its own, with ice or with tonic.

Ingredients
1 bottle of gin
275 g (10 oz) sugar
Ripe diddle-dee berries

TEABERRY VODKA

You will need a spare empty bottle to divide the vodka between the two bottles.

1. Empty the vodka into a jug.
2. Divide the teaberries between each bottle ensuring there is a depth of at least 5–7 cm.
3. Divide the sugar between the two bottles before topping up each bottle with the vodka.
4. Shake the bottles well and place in a dark cupboard for around 3–4 months.
5. Shake the bottles every now and again to ensure the juice from the teaberries is flavouring the vodka well. Again, the longer you can resist the temptation to sample the product the better!
6. When drinking, separate the fruit from the vodka using a tea strainer and return the berries to the bottle. You can drink this on it's own, with ice or with tonic or lemonade.

Ingredients
1 bottle of vodka
125 g (6 oz) sugar
Teaberries

Index

Checklist

This checklist covers all the plants illustrated in this book.

FLOWERING PLANTS		DATE	LOCATION
Astelia (Soft-camp-bog)	*Astelia pumila*		
Bead Plant	*Nertera granadensis*		
Arrow-leaved Marigold	*Caltha sagittata*		
Berry-lobelia	*Pratia repens*		
Balsam Bog	*Bolax gummifera*		
Wild Celery	*Apium australe*		
Diddle-dee	*Empetrum rubrum*		
Buttonweed	*Leptinella scariosa*		
Christmas Bush	*Baccharis magellanica*		
Sea Cabbage	*Senecio candidans*		
Marsh Daisy	*Aster vahlii*		
Vanilla Daisy	*Leucheria suaveolens*		
Fachine	*Chiliotrichum diffusum*		
Groundsel	*Senecio vulgaris*		
Antarctic Hawkweed	*Hieracium antarcticum*		
Orange Hawkweed	*Hieracium aurantiacum*		
Falkland Lavender	*Perezia recurvata*		
Coastal Nassauvia	*Nassauvia guadichaudii*		
Falkland Smooth Ragwort	*Senecio vaginatus*		
Falkland Woolly Ragwort	*Senecio littoralis*		
Sheep's Sorrel	*Rumex acetosella*		
Boxwood	*Hebe elliptica*		
Lady's Slipper	*Calceolaria fothergillii*		
Yellow Lady's Slipper	*Calceolaria biflora*		
Felwort	*Gentianella magellanica*		
Pig Vine	*Gunnera magellanica*		
Mountain Berry	*Pernettya pumila*		
Pale Maiden	*Olsynium filifolium*		

Yellow Pale Maiden	*Sisyrinchium chilense*		
Teaberry	*Myrteola nummularia*		
Dog Orchid	*Codonorchis lessonii*		
Gaudichaud's Orchid	*Chloraea gaudichaudii*		
Pale Yellow Orchid	*Gavilea australis*		
Yellow Orchid	*Gavilea littoralis*		
Gorse	*Ulex europaeus*		
Almond Flower	*Luzuriaga marginata*		
Field Mouse-ear	*Cerastium arvense*		
Dusty Miller	*Primula magellanica*		
Pimpernel	*Anagallis alternifolia*		
Falkland Strawberry	*Rubus geoides*		
Prickly Burr	*Acaena magellanica*		
Stonecrop	*Crassula moschata*		
Sundew	*Drosera uniflora*		
Falkland Thrift	*Armeria macloviana*		
Common Violet	*Viola maculata*		
Water-milfoil (Pondweed)	*Myriophyllum quitense*		
Scurvy Grass	*Oxalis enneaphylla*		
TREES		DATE	LOCATION
Monterey Cypress	*Cupressus macrocarpa*		
GRASSES AND RUSHES		DATE	LOCATION
Cinnamon Grass	*Hierochloe redolens*		
Tussac Grass	*Poa flabellata*		
White Grass	*Cortaderia pilosa*		
Native Woodrush	*Luzula alopecurus*		
Short Rush	*Rostkovia magellanica*		
Tall Rush	*Marsippospermum grandiflorum*		
FERNS		DATE	LOCATION
Tall Fern	*Blechnum magellanicum*		
Small Fern	*Blechnum penna-marina*		

Falklands Conservation has been protecting the wildlife of the Falkland Islands for over 25 years. We are a thriving Falklands-based charity at the heart of the Islands' community.

Falklands Conservation has in the past supported and run projects to record and catalogue the plants of the Islands. This work resulted in a number of publications including *Wild Flowers of the Falkland Islands* (T H Davies and J H McAdam,1989), *Flowering Plants of the Falkland Islands* (Robin Woods, 2002) both of which are now out of print. *The Vascular Flora of the Falklands Islands: An Annotated Checklist and Atlas,* also published in 2002 (David A. Broughton and James H McAdam), provides an invaluable baseline on the status and distribution of Falkland plants, vital for their effective conservation and protection.

In 2007 a new plants programme is focused on identifying Important Plant Areas, producing Species Action Plans for rare species or those under threat and developing sustainable land-use practices. This is in conjunction with the Falkland Islands Government Department of Agriculture.

The Falkland Islands are small and remote with limited resources but with hugely important wildlife to protect. Falklands Conservation is heavily dependent on public support to fund its work. You can help us by becoming a member or by making a donation. We welcome visitors to our offices and to the National Herbarium in Stanley, or you my find our more about us on our website:

www.falklandsconservation.com